ESCAPING WATER

Written by
Read With You Science Academy

Read With You

The characters and events portrayed in this book are fictitious. Any similarity to real persons, living or dead, is coincidental and not intended by the author.

Published by Read With You Publishing
Designed by Read With You Science Academy
Read With You and associated logos are trademarks and/or registered trademarks of Read With You L.L.C.

ISBN-13: 978-1-64616-111-9
ISBN-10: 1-64616-111-4

Third Edition April 2021
Printed in the United States of America.

Welcome to the

READ WITH YOU SCIENCE ACADEMY!

Today we will do an experiment called

ESCAPING WATER

Before we begin, let's review some important words:

wick:
A string or piece of material that is usually used in a candle.

to absorb:
To take in (something, such as a liquid) in a natural or gradual way.

to transfer:
To move (someone or something) from one place to another.

Remember:
Be a careful scientist! Clean up any water that spills right away!

QUESTION:
What I want to know.

Can water move from one glass to another without being poured?

- [ ] Yes, water can move from one glass to another without being poured.
- [ ] No, water cannot move from one glass to another without being poured.

HYPOTHESIS:

What I think will happen and why.

I think water can move to from one glass to another without being poured because...

I think water cannot move from one glass to another without being poured because...

Water **can** transfer from one glass to another because:	Water **cannot** transfer from one glass to another because:
The paper towel "wick" will absorb and transfer the water from one glass to the other.	The paper towel "wick" will not absorb and transfer the water.
The water is not solid and can move easily.	It is impossible to move liquid without pouring it.

What do you think?
Circle your hypothesis!

EXPERIMENT:
How we will test our hypothesis.

THINGS WE NEED

- A glass of water
- An empty glass
- Some paper towels
- Food coloring
- A shallow bowl or tray to catch any water that may spill
- A watch or a stopwatch

water

empty glass

watch

paper towels

food coloring

PROCEDURE:

The steps to complete the experiment.

Step 1A

Twist two sheets of paper towel together until they form something that looks like a piece of rope. This is the "wick" that will **absorb** and transfer the water.

Step 2A

- Fill one glass with water and place it in the shallow bowl or tray. Place the empty glass next to the filled glass. They should be as close together as possible.
- Put 3 drops of food coloring in the filled glass.

Step 3A

Place one end of the paper towel wick into the glass filled with water and the other into the empty glass.

- Start your timer.
- Then, **observe** what happens. (This experiment may take a little time and patience.)
- Remember: Use a towel or paper towel to clean any water on the table or floor!

Step 4A

- When the water starts to fill the empty glass, you may stop timing.
- Record the time and what you observed in your science journal.

Write down what you see

In your science journal,

write your results!

- What was your hypothesis? (Look on page 5 of this book!)
- Was your hypothesis correct?
- What happened to the water?

Let's do it again!

How will different levels of water affect the transfer time? Let's do an experiment to find out. You will need another paper towel wick, one half-filled glass of water, an empty glass, and a watch or a stopwatch to time the transfer.

QUESTION:

What I want to know.

Will the water transfer faster from a half-filled glass?

HYPOTHESIS:

What I think will happen and why.

I **think** the transfer will be faster because…

I **don't think** the transfer will be faster because...

What do you think?

Write your new hypothesis in your science journal.

PROCEDURE:

The steps to complete the experiment.

Step 1B

- Twist paper towels together until they form something that looks like a piece of rope. This is the “wick” that will absorb and **transfer** the water.

Step 2B

- Fill one glass halfway with water and place it in the shallow bowl or tray. Place the empty glass next to the filled glass. They should be as close together as possible.
- Put 3 drops of food coloring in the half-filled glass.

Step 3B

- Place one end of the paper towel wick into the half-filled glass of water and the other into the empty glass.
- Start the timer.
- Remember: Use a towel or paper towel to clean any water on the table or floor!

Step 4B

- When the water starts to fill the empty glass, you may stop timing.
- Record the time in your science journal.

Write down what you see

In your science journal,

write your results!

- What was your hypothesis? (Look on page 11 of this book!)
- Was your hypothesis correct?
- How much time did it take for the water to transfer from one glass to the other?

(Optional) Draw a **diagram** of what happened during this experiment.

Why did this work?

- In this experiment, you saw water from the full glass transfer to the empty glass using the paper towel as a "wick." The paper towel absorbed the water from one glass and emptied it into the other glass until both glasses contained equal amounts of liquid.
- The experiment worked because of something called **capillary action**. This is the process of water moving from one place to another through tiny gaps or tubes in a material.
- The paper towel "wick" you made allowed the water to move using capillary action because there are tiny gaps in the material. The water from the first glass "escaped" to the empty glass, and you were able to move water from one glass to another without pouring it!

SCIENCE

- Three forces work together in the process capillary action: **cohesion**, **adhesion**, and surface tension.
- Water molecules are cohesive and they like to connect to one another. Molecules on the surface have no neighboring molecules above to connect to, so they are strongly attached to their nearest neighboring molecules above and to the side of them. **Surface tension** is a product of the strong interactions between water molecules.
- As we saw in the experiment, the water molecules wanted to **adhere**, or stick, to the paper towel "wick" more than they wanted to stay connected in the glass. The surface tension tried to keep the surface intact, but the adhesion to the paper towel was stronger than the cohesive forces between the water molecules. As one water molecule moved up the paper towel, it pulled the other molecules along with it.
- Capillary action is how water moves from a plant's roots to its other parts.

GLOSSARY

To absorb: To take something in, little by little.

To adhere: To stick or attach to something.

Adhesion: The act of sticking or attaching to something.

Capillary action: The process of water moving from one place to another through tiny gaps or tubes in a material.

Cohesion: To be closely united.

Diagram: A drawing that explains or shows the parts of something.

To observe: To watch something carefully.

Surface tension: The force that causes the molecules on the surface of a liquid to be pushed together and form a layer.

To transfer: To move (someone or something) from one place to another.

NEW IDEAS

How could we change this experiment?

- Could you use a piece of yarn as the wick in place of the paper towels?
- Can you think of any other material that could be used?

Can you think of any ideas to improve the experiment?

- Can you compare the transfer rate by using other liquids?

JOURNAL

Made in the USA
Monee, IL
05 November 2026

67386145R00017